AF228280

SAVING THE SNOW LEOPARD

LOUISE SPILSBURY

CHERITON
CHILDREN'S BOOKS

Published in 2023 by **Cheriton Children's Books**
PO Box 7258, Bridgnorth WV16 9ET, UK

© 2023 Cheriton Children's Books

First Edition

Author: Louise Spilsbury
Designer: Paul Myerscough
Editor: Jane Brooke
Proofreader: Tracey Kelly
Consultant: David Hawksett, BSc

Picture credits: Cover: Shutterstock/Warren Metcalf. Inside: p1: Shutterstock/Sergei Primakov; pp4-5: Shutterstock/Abeselom Zerit; pp6-7: Shutterstock/Abeselom Zerit; p7t: Shutterstock/Eric Isselee; p7l: Shutterstock/BearFotos; p7r: Shutterstock/Eric Isselee; p7b: Shutterstock/Eric Isselee; p9c: Shutterstock/Asmakhan992; pp: Shutterstock/Abeselom Zerit; pp: Shutterstock/Slowmotiongli; p12b: Shutterstock/Adalbert Dragon; p13b: Shutterstock/USBFCO; p13r: Shutterstock/Slowmotiongli; pp14-15: Shutterstock/Dr Ajay Kumar Singh; pp16-17: Shutterstock/Warren Metcalf; p17t: Shutterstock/Chatursunil; p17b: Shutterstock/Chatursunil; pp18-19: Shutterstock/Nora Yusuf; p19c: Shutterstock/Bildagentur Zoonar GmbH; pp20-21: Shutterstock/Wang LiQiang; p21t: Shutterstock/Mikhail Semenov; p21c: Shutterstock/Ondrej Prosicky; p21b: Shutterstock/Belovodchenko Anton; pp22-23: Shutterstock/Dennis W Donohue; p23t: Shutterstock/Africa Studio; p23b: Shutterstock/Africa Studio; pp24-25: Shutterstock/Sergei Primakov; pp26-27: Shutterstock/Ondrej Chvatal; p27t: Shutterstock/Bildagentur Zoonar GmbH; p27c: Shutterstock/Kwadrat; p27b: Shutterstock/Belizar; pp28-29: Shutterstock/Kwadrat; p29t: Shutterstock/NadyGinzburg; p29c: Shutterstock/Krakenimages.com; p29b: Shutterstock/Krakenimages.com; pp30-31: Shutterstock/Vincent Legrand; p30t: Shutterstock/Paul Gibbings; pp32-33: Shutterstock/Jim Cumming; p33t: Shutterstock/Dennis W Donohue; p33c: Shutterstock/Holly S Cannon; p33b: Shutterstock/Quincy Floyd; pp34-35: Shutterstock/SofotoCool; p35t: Shutterstock/Belizar; p35c: Shutterstock/Sergey Novikov; p35b: Shutterstock/Sergey Novikov; pp36-37: Shutterstock/Lauren Bilboe; p36b: Shutterstock/Andreas Rose; pp38-39: Shutterstock/Jo Reason; p38b: Shutterstock/Slowmotiongli; p39c: Shutterstock/The Len; p39b: Shutterstock/PhotocechCZ; pp40-41: Shutterstock/Kwadrat; p40c: Shutterstock/Pixelheadphoto Digitalskillet; p40b: Shutterstock/Pixelheadphoto Digitalskillet; p42t: Shutterstock/Vladimir Turkenich; p43b: Shutterstock/LeManna; p44t: Shutterstock/Panumas Yanuthai; p45b: Shutterstock/Chrisdorney.

Printed in the United States

Publisher's Note: The information in the Kids on a Mission features in this book are suggestions for actions that children can take to help protect endangered animals, based on extensive research by the author and consultant. The email addresses and the children featured in the photographs are for illustrative purposes only.

Please visit our website,
www.cheritonchildrensbooks.com,
to see more of our high-quality books.

CONTENTS

SNOW LEOPARDS IN DANGER

Snow leopards are truly amazing animals. These beautiful beasts are known as "mountain ghosts." That is because they are so secretive and rarely seen. Sadly, the magnificent big cats are in danger. And they're in danger because of people.

IN DANGER AND DISAPPEARING

Today, around only 3,500 to 7,000 snow leopards live in the wild. That is because of human actions. Snow leopards have been hunted in huge numbers by people. Hundreds of snow leopards are still killed every year. They are killed for their skin and body parts. People have also taken over snow leopard land to make room for **mines**. They have built homes on the land, too. Livestock, or farm animals, are kept on the land. That leaves less food for the wild animals that snow leopards eat. Those animals are disappearing, leaving snow leopards hungry. When the hungry snow leopards then eat farm animals, people kill them.

The International Union for **Conservation** of Nature (IUCN) keeps a record of the world's **species** and how at risk of **extinction** they are. It is called the **Red List.**

There are more than **142,500** species on the Red List.

The snow leopard is listed as **vulnerable.**

NO MORE SNOW LEOPARDS?

No one knows exactly how many snow leopards exist. That is because the animals are so secretive. They also live high on mountains, where few people go. However, **conservationists** are warning us that we need to do more to save the remaining snow leopards on Earth. The number of snow leopards is decreasing every year. If we do not do something, one day they may be gone forever.

HELP THE SNOW LEOPARD!

There is still hope for the snow leopard, and it is not too late to save it. People everywhere have heard the snow leopard's cry for help. And they are making it their mission to help these amazing animals survive. In this book, we'll learn about the snow leopard and why it is in danger. We'll discover what people are doing to help snow leopards and how they have built a career in conservation. We'll find out how kids everywhere can make it their mission to help save the snow leopard. And we'll learn how you can make a career in conservation your mission. Feeling mission-ready? Then read on!

▲ Beautiful snow leopards are disappearing fast. It would be a tragedy if these incredible animals are lost forever.

"The snow leopard is absolutely magnificent. It represents really what **endangered** species are all about."

Jack Hanna, zoo director and animal expert

CHAPTER ONE

MEET THE SNOW LEOPARD

Snow leopards are known as "ghost cats" because they are so hard to find. The coloring of their coats makes them difficult to see against the rocky mountains where they live.

SPECIAL SPOTS

A snow leopard has gray-black spots on its pale-colored fur. The pattern helps snow leopards blend in against the rocky slopes on which they hunt. Every snow leopard has its own special pattern of dark spots. That helps scientists and conservationists identify each individual animal.

LONG LEOPARD

From its nose to its tail, an adult snow leopard measures around 7 feet (2.1 m) long. A snow leopard's tail alone can measure 3 feet (1 m) long! The snow leopard has short, round ears. It has a wide, short nose that warms the air before it reaches the cat's lungs. The snow leopard uses its long, powerful back legs to pounce on **prey**.

SNEAKY HUNTER

A snow leopard has huge paws with fur on the bottom. At the end of the paws are very sharp, long claws. A snow leopard can pull in its claws to keep them from scraping against rocks. That helps it to move silently to sneak up on prey. A snow leopard also has very sharp teeth called fangs.

In a Group

Scientists group animals to help them classify, or order, them. Snow leopards belong to a group of animals called *Panthera*. Other big cats are also members of the group. They are lions, leopards, jaguars, and tigers. Interestingly, the snow leopard is the only one of the big cats that cannot roar!

WHERE SNOW LEOPARDS LIVE

Snow leopards live high in the steep, rocky mountains of central Asia. These mountains are spread across 12 different countries. Today, snow leopards are found in just a small part of the **territory** in which they once lived. The total area of land over which snow leopards are found covers 1.2 million square miles (2 million sq km). That is about the same size as Greenland or Mexico. In the past, snow leopards had 6.6 million square miles (10.47 million sq km) of land on which to roam. That means that their **habitat** has reduced by about 80 percent.

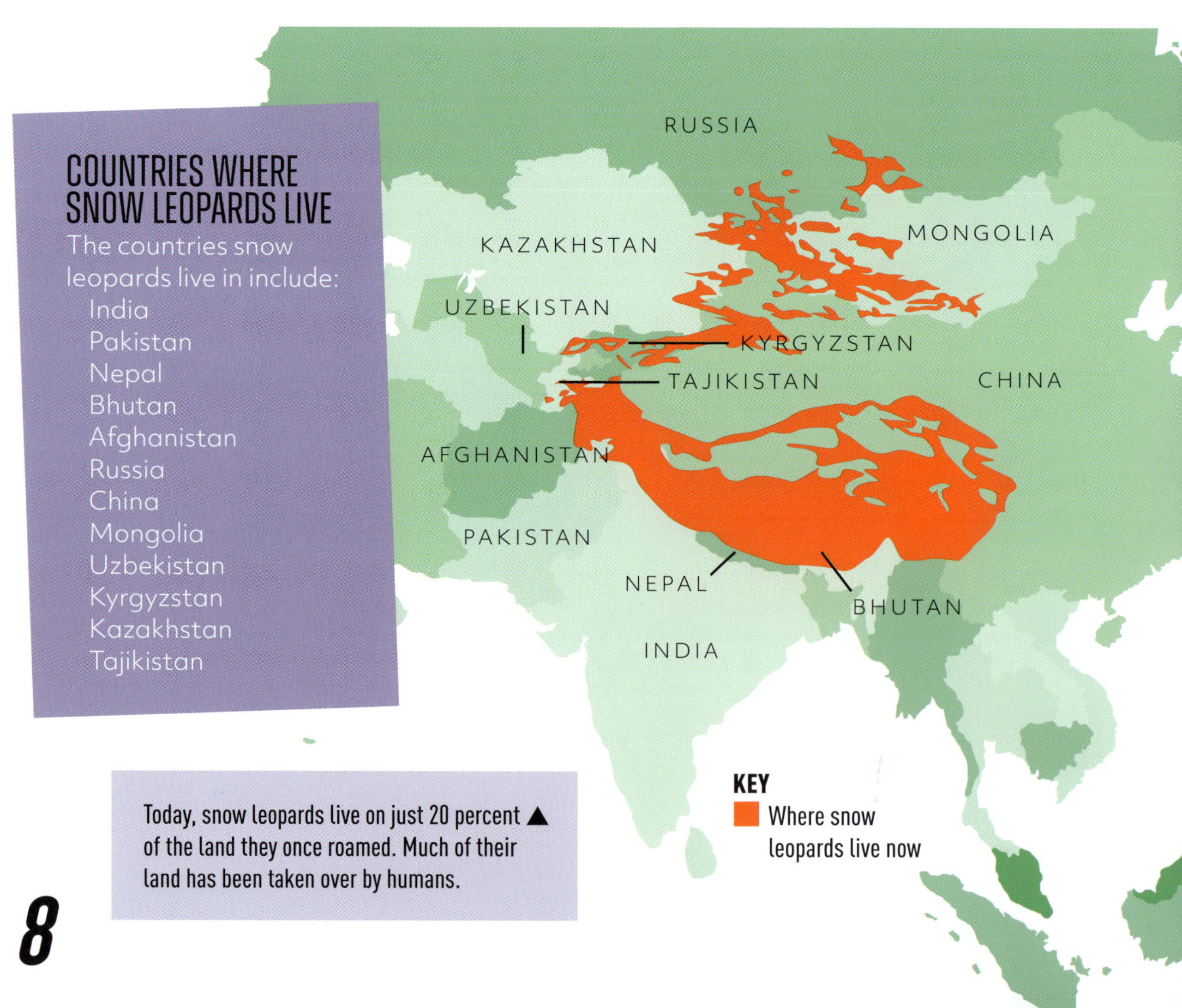

Today, snow leopards live on just 20 percent ▲ of the land they once roamed. Much of their land has been taken over by humans.

LIVING IN THE MOUNTAINS

Snow leopards have physical features that help them survive in some of the harshest conditions on Earth. They have the longest and thickest fur of any big cat. That keeps them warm. They can wrap their long, thick tail around themselves to keep out the cold, too. The wide footpads on their paws are covered with a cushion of fur. That fur protects the pads from cold snow and rough rocks. It also helps spread their weight more evenly over the snow. That keeps them from sinking into it.

▼ Snow leopards live in steep, rocky, and high mountains in Asia that are often covered in snow.

SURVIVING ON THE SLOPES

Snow leopards live in places about 9,800 to 14,800 feet (3,000 to 4,500 m) above **sea level**. They keep mainly to cliffs and rocky slopes. They have short, powerful front legs and strong chest muscles to help them climb. They usually hunt between dusk and dawn. They can travel more than 25 miles (40 km) in one night when hunting. If the amount of land they can hunt on keeps declining, they will starve.

WHY SHOULD WE HELP THE SNOW LEOPARD?

Snow leopards are incredibly important to the **environments** in which they live. Snow leopards are apex **predators**. That means they are at the top of their **food chain**. They hunt other animals in the food chain. Snow leopards have no natural predators. The animals help control their ecosystem. That is because they hunt animals within it that would otherwise become too great in number. Animals that snow leopards hunt are herbivores. These are plant-eating animals. If they become too numerous, they could eat too many of the plants they feed on. These plants would then die out. That would leave no food for the herbivores. They, too, would die.

Snow leopards help keep the ▶ habitats they live in healthy. Without snow leopards, there would not be enough plant foods for herbivores to eat.

KEEPING HABITATS HEALTHY

By protecting snow leopard habitats, we are also protecting the habitats for all the other plants and animals that live there. Many different **organisms** live in the mountains where snow leopards are found. If these places are kept healthy for snow leopards, they are kept healthy for every other organism that lives there, too.

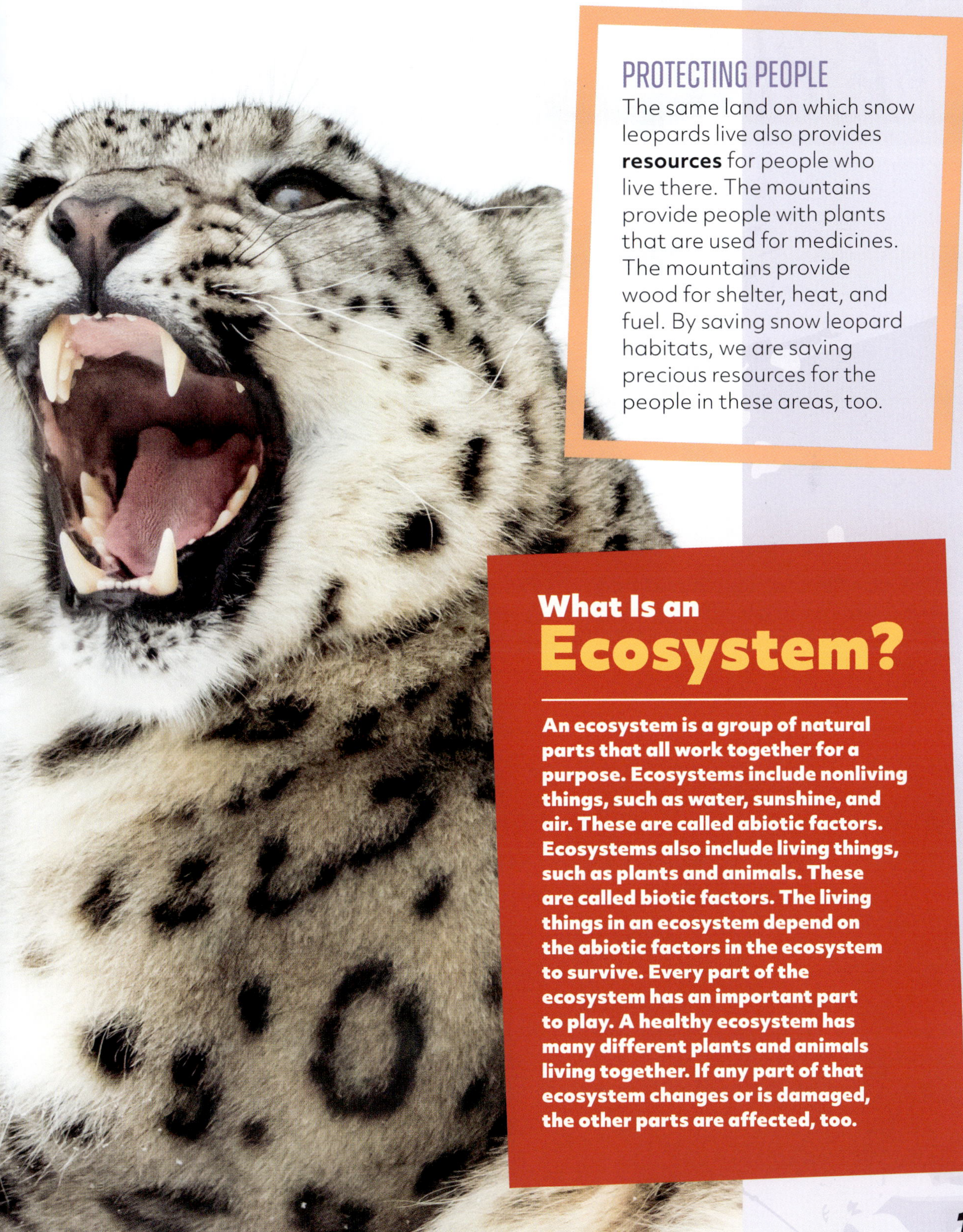

The same land on which snow leopards live also provides **resources** for people who live there. The mountains provide people with plants that are used for medicines. The mountains provide wood for shelter, heat, and fuel. By saving snow leopard habitats, we are saving precious resources for the people in these areas, too.

What Is an Ecosystem?

An ecosystem is a group of natural parts that all work together for a purpose. Ecosystems include nonliving things, such as water, sunshine, and air. These are called abiotic factors. Ecosystems also include living things, such as plants and animals. These are called biotic factors. The living things in an ecosystem depend on the abiotic factors in the ecosystem to survive. Every part of the ecosystem has an important part to play. A healthy ecosystem has many different plants and animals living together. If any part of that ecosystem changes or is damaged, the other parts are affected, too.

DESIGNED TO HUNT

Snow leopards are amazing athletes and acrobats. They can make huge leaps over deep holes in the ground called ravines. They can race across steep, uneven rocks to catch prey. They are one of nature's killing machines.

NIGHTTIME HUNTING

Like all cats, snow leopards can see well in the dark. That makes them deadly nighttime hunters. Their short, rounded ears can also hear animals moving from far away.

SECRET STALKER

The snow leopard quietly stalks its prey and stays out of sight. When it is close to the prey, it jumps out and chases the animal down. Then it sinks its long, deadly fangs into the victim.

SUPERFAST SNOW LEOPARD

If a snow leopard needs to chase its prey, it can run at breathtaking speed. Its long, strong tail helps balance the animal's body as it races over steep, uneven ground.

POUNCE POWER

Snow leopards like to surprise their prey by attacking from above. They have long, powerful back legs, and they can travel as far as 30 feet (9 m) in one leap. That's six times their body length!

FAVORITE FOODS

A snow leopard can kill prey two to three times its own weight! A snow leopard's favorite foods are wild goats and sheep. The cat eats ibex, which is a wild goat. It also eats argali, which is a wild sheep. Blue sheep are also on the menu. The snow leopard eats marmots and other **rodents**, too.

13

PEOPLE ON A MISSION

There are people all over the world who are working to save the snow leopard. They have made it their mission to help the leopard survive. Many work in conservation to protect this special species from the dangers it faces.

PEOPLE ON THE GROUND

Some people work to protect the snow leopard's habitat. Some conservationists work with **Geographic Information Systems (GIS)** to map snow leopard habitat. Conservation scientists study snow leopards in the wild. They learn what the animals need to survive. Conservationists also work with local people to help them save the snow leopard and its habitat. Ecotourism conservationists work in **sustainable** tourism. The money they raise helps maintain the snow leopard's habitat.

PEOPLE IN OFFICES

People in conservation organizations work with governments in countries in which snow leopards live. Together, they plan protected areas and corridors of land that connect snow leopard populations. **Policy** and **advocacy** conservationists help countries make laws that protect the snow leopard. There are also charities in which people work hard to raise money to save snow leopards. **Communications** and public relations experts help organizations raise awareness about snow leopards.

Some scientists who help snow leopards work in zoos or labs. Zoologists and wildlife technicians study snow leopards to find out more about them and ways we can help them. They also study data, or information, other conservationists collect.

▼ Conservationists brave freezing snow and ice to capture photographic evidence that shows where snow leopards live.

MAKE IT YOUR MISSION

You can help snow leopards and other endangered animals by taking action and planning a career in conservation. Here's how:

1. In this book, you'll discover what actions kids on a mission can take to help snow leopards. Use them to inspire your own actions to rescue snow leopards.
2. You'll also discover some of the careers that people on a mission have in snow leopard conservation. As you read about each one, think about whether that career in conservation might suit you.
3. At the end of this book, you'll find a guide to how to build a career in conservation. Check it out to discover how you can make saving animals your life mission.

PEOPLE EVERYWHERE

Ordinary people everywhere, including kids just like you, raise money to save snow leopards. They share the information they learn about the threats snow leopards face with their friends and family. They join organizations that help snow leopards.

CHAPTER TWO

HOMES UNDER THREAT

Snow leopard habitat is under threat. Snow leopards have lost a huge amount of the land on which they once lived, and more is lost every year. Snow leopards have already disappeared from some areas where they once lived. That includes parts of Mongolia, in Asia. How will they cope if their habitat continues to shrink?

TAKEN OVER BY PEOPLE

Snow leopards need huge areas of land to thrive. They travel long distances to find enough food to survive. Unfortunately, even the harsh mountains where they roam are being taken over by people. Humans are building fences and barriers on the land. They are building roads and railroads. They are breaking up the habitats where snow leopards live. That reduces the amount of land across which the animals can roam.

DESTROYING SNOW LEOPARD HOMES

Snow leopard territory is also being taken over by industry. People are building dams for **hydroelectric power**. They are setting up pipelines for extracting, or taking out, resources such as oil and natural gas. They are digging mines to extract copper, gold, and other **minerals**. Mining companies use dynamite to blast open large pits in the land. That damages the landscape. The mines also use deadly chemicals to extract **precious metals**. Those chemicals run off from the mining sites. The chemicals run into surrounding land and water. They then **pollute** water that snow leopards drink. Large areas of snow leopard land are being destroyed due to these activities.

KEEPING SNOW LEOPARDS SAFE

To help protect snow leopards, conservationists have set up wildlife reserves. These are areas in which snow leopards should be safe from threats such as mining and road building. Wildlife reserves are monitored by people working in snow leopard conservation. They try to keep the snow leopards in the reserves safe. Conservationists also create wildlife corridors for the snow leopards. They are stretches of land that connect two wildlife reserves. The corridors give snow leopards the chance to move safely between important habitats.

◀ Humans are spreading farther and deeper into snow leopard habitat. That is putting snow leopards at risk.

"For conservation measures to succeed, they have to work for local people, too, so humans and snow leopards can successfully coexist."

Jonathan Hanson, leader of a study into snow leopard conservation

SPOTLIGHT ON *FARMING*

Farming is a major problem for snow leopards. Farmers clear snow leopard territory so they can raise livestock on it. That means there is less space for snow leopards and the prey animals on which they usually feed. Today, there is around 10 times more livestock than wild prey in some snow leopard areas.

HUNGRY HUNTERS

When there are fewer wild animals for hungry snow leopards to eat, they look elsewhere for food. That often includes farm animals. Many farmers keep their livestock inside corrals. They are **enclosures** with low stone walls over which snow leopards can easily jump. Once a snow leopard gets into the corral, the livestock panic and run around. That triggers the snow leopard's killing instinct. It may then kill many of the animals.

POOR FARMERS

Many of the herders who use the mountains of central Asia to raise livestock are very poor. If a snow leopard kills most of a farmer's livestock, their family may be left with nothing. It is thought that more than half of all snow leopards are killed by angry farmers. They kill the animals because they eat the farmers' livestock.

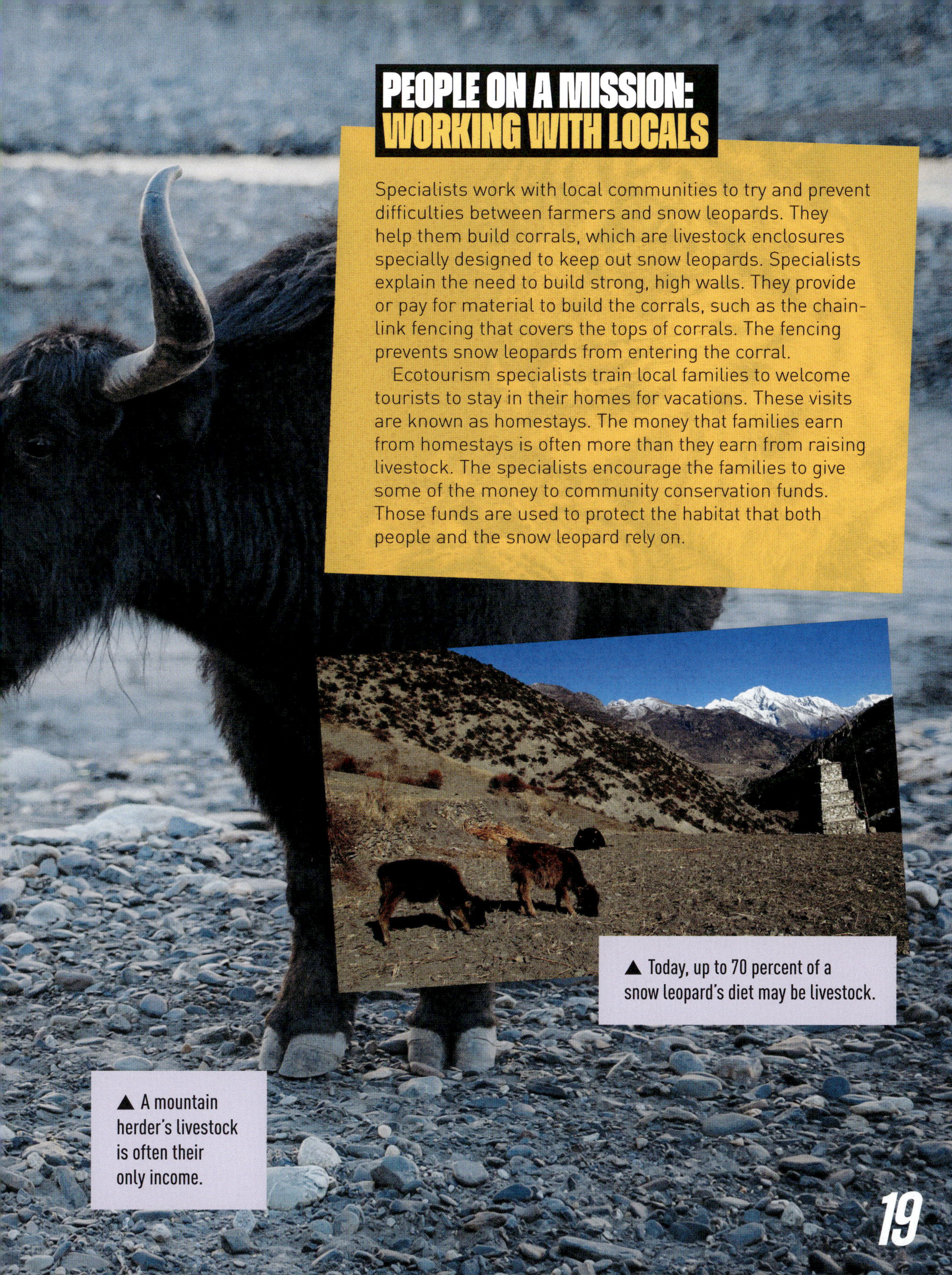

Specialists work with local communities to try and prevent difficulties between farmers and snow leopards. They help them build corrals, which are livestock enclosures specially designed to keep out snow leopards. Specialists explain the need to build strong, high walls. They provide or pay for material to build the corrals, such as the chain-link fencing that covers the tops of corrals. The fencing prevents snow leopards from entering the corral.

Ecotourism specialists train local families to welcome tourists to stay in their homes for vacations. These visits are known as homestays. The money that families earn from homestays is often more than they earn from raising livestock. The specialists encourage the families to give some of the money to community conservation funds. Those funds are used to protect the habitat that both people and the snow leopard rely on.

▲ Today, up to 70 percent of a snow leopard's diet may be livestock.

▲ A mountain herder's livestock is often their only income.

LIVING WITH THE LEOPARD

Many other animals are found in the high, rocky mountains of Asia where snow leopards live. Sadly, many of these animals are endangered, too.

THE SAKER FALCON

The saker falcon is a large, strong **bird of prey**. It has big feet and pointed wings. These birds can be found in Mongolia, southern Siberia, and in Russia. They have large eyes and a short, hooked beak. They make their nests on rock faces. Today, saker falcons are extremely rare.

THE MARKHOR GOAT

The markhor is also known as the screw-horned goat. This large, wild goat was once found from Kashmir to Afghanistan. It is now greatly reduced in number. Its coat is reddish-brown in summer and long and gray in winter. Males have a long, heavy fringe on their throat and chest.

THE MANUL

The manul is a sulky-looking cat! It is about the same size as a pet cat. It has a thick body, covered with soft fur. The dark, woolly fur on its belly is twice as long as the rest of its fur. The fur coat keeps the animal warm. Its fluffy tail can be wrapped around its body like a shawl. Manuls hide for much of the day in caves or hollows under stones. They hunt at night.

THE ARGALI

The argali is the largest wild sheep in the world. It is best known for its huge, spiral-shaped horns. The argali can grow up to 6 feet (1.8 m) long and can weigh 59 pounds (27 kg). Argali live in remote, or faraway, places alongside the snow leopard, so we don't know how many exist. We do know that argali are endangered, and their numbers are dropping.

CHAPTER THREE

UNDER THREAT FROM HUNTING

Poaching is the illegal killing or capture of animals. Poachers hunt and kill snow leopards for their body parts, which they sell. The parts include fur, skin, and bones. They also include claws and teeth. The capture and killing of snow leopards for the illegal wildlife **trade** is a major threat to the animals' survival.

KILLED FOR ITS FUR

The beautiful smoky gray fur of the snow leopard is unique among wild cats. It is the fur that first put snow leopards in danger. In the eighteenth century, snow leopard fur became very popular. It was first sold in Europe and then worldwide. By the early twentieth century, around 1,000 snow leopard skins were traded around the world each year. As a result, the number of snow leopards drastically dropped. In 1975, snow leopards were included on the list of species threatened with extinction by the Convention on International Trade of Endangered Species (CITES). The trade in snow leopards was then banned. Despite that, illegal hunting of snow leopards still continues.

DRIVEN TO POACH

Although it is illegal to trade in snow leopards or their body parts, people who hunt snow leopards are often very poor. They turn to poaching to feed their families. They can make enough money from selling a snow leopard to feed their family for a year. The snow leopard's fur is usually made into rugs. Some snow leopards are sold for taxidermy. That is the process of stuffing a dead animal and then putting it on display.

SPOTLIGHT ON POACHING

Many of the people who live near snow leopard homes have beliefs that do not allow them to harm nature. Unfortunately, the high value of snow leopard fur and body parts can prove too tempting to people who are very poor.

This ranger is installing an infrared ▶ camera. Wildlife rangers face dangers in their jobs. Wild animals are a threat. So, too, are poachers, who can react badly if caught red-handed.

A CRUEL DEATH

Hunters mostly catch and kill snow leopards using guns and traps. The traps may be steel jaw traps. They may be loops of wire called snares. Hunters also kill snow leopards by hitting them with clubs. They poison them, too. Snow leopards caught in traps often die slowly. So, the poacher may shoot the animal or club it to death while it is caught in the trap.

MORE POACHER PROBLEMS

Snow leopards also die in traps that poachers set to catch smaller endangered animals. Sometimes, poachers even set thousands of these traps at a time. Snow leopards are often caught in these traps. Their remains may not be found until years later. Poachers who catch smaller animals cause another problem for snow leopards. Without healthy populations of prey animals, snow leopards cannot survive.

PEOPLE ON A MISSION: STOPPING LEOPARD POACHERS

Wildlife rangers work in different ways to protect the snow leopard from poaching. They work with local people to learn where snow leopards have been spotted. Local people may also have information about where poachers set their traps. Wildlife officers find and remove traps and wire snares. If they find a poacher, they have the power to arrest them.

Wildlife rangers set up camera traps. They have motion sensors. These are devices that sense movement. The cameras start filming only when an animal walks into the frame. They can even film at night, thanks to their infrared sensors. The sensors detect heat given off by organisms. The cameras help rangers track changes in the snow leopard population. They also help rangers identify poachers who are caught on camera. Rangers may fit radio collars to some snow leopards. The **Global Positioning System (GPS)** collars transmit, or send out, the location of the animals. That gives rangers information about the snow leopard's habitat, behavior, and movements. That data helps scientists figure out where to find the animals and how rangers can help keep them safe.

Rangers persuade local people to stop setting poaching traps. They explain that without the snow leopard, there would be too many hoofed animals feeding on the mountain plants. That would reduce the area to empty, rocky peaks. That would affect wildlife and trees. It would affect water, too. People need all those resources, so protecting the snow leopard is in their interest. Rangers also pay local people to capture photos of snow leopards rather than killing them.

BABIES UNDER THREAT, TOO

It is not just adult leopards that are threatened by poaching: so too, are babies. Female snow leopards care for their cubs alone, so if a mother dies her cubs will almost certainly die.

A SAFE PLACE

A snow leopard mother is pregnant for around 93 to 110 days before she gives birth to her cubs. Before the babies are born, she finds a safe place to give birth. That might be beneath rocks or in a hole in a rock. She lines the den with fur from her belly. Snow leopard mothers have between one and three cubs.

CARING FOR HER CUBS

Snow leopard cubs have a coat of thick fur when they are born. They are small and helpless. They don't even open their eyes until they are seven days old. A snow leopard mother has to do everything she can to keep her babies alive. When the cubs are newborn, the mother **nurses** her babies day and night.

CUBS IN DANGER

The mother snow leopard leaves the cubs in the den only to hunt for food. The cubs are more likely to be safe from danger in the den. Predators, such as wolves and people, are a risk to snow leopard cubs.

LEOPARD LEARNING

While they are in the den, snow leopard cubs play with their brothers and sisters. That helps them build strength. It also teachers them to pounce and bite. When cubs are a few months old, they leave the safety of their den. They follow their mom as she hunts for prey. They live with her until they are between 18 and 22 months old.

CLIMATE CHANGE THREATS

Climate change is now affecting everyone and everything on Earth. It is causing our planet to get warmer. That is causing snow to melt, which is affecting humans and wildlife such as snow leopards.

▲ Snow leopards live in only one type of habitat. They cannot move to a new type if their homes are lost due to climate change.

"Urgent action is needed to curb climate change ... otherwise, the ghost of the mountains could vanish ..."

Rishi Kumar Sharma, WWF Global Snow Leopard Leader

WHAT IS CLIMATE CHANGE?

Climate change is a change in temperature or weather in an area over a long period of time. Climate change has always happened slowly throughout Earth's history, due to natural factors such as changes in the sun and **geological** activity. However, the climate change we are experiencing today is largely caused by human activities. They include burning **fossil fuels**, which releases gases into Earth's **atmosphere**. These gases circle the planet like a blanket, causing it to heat up.

▲ Gasoline is a fossil fuel that is burned by cars to power them. During the process, harmful gases are relased into the atmosphere.

CHANGING HABITATS

Snow leopards live in high, steep, and rocky places. There, few plants grow. Snow leopards prefer to live above the tree line. The tree line is the highest point on a mountain that is warm enough for thick forests to grow. Climate change is causing temperatures to rise across the mountains of central Asia. In some areas of snow leopard habitat, it has become 3 degrees hotter in the last 20 years. That means the tree line is getting higher and thick forests will grow higher up the mountains. That will reduce the areas of open land where the snow leopard can live and hunt.

SPOTLIGHT ON WARMING MOUNTAINS

Snow leopards are highly adapted to their homes in cold, high mountains. That means their bodies and behavior patterns have changed to cope with their environment. So warming mountains are a major problem for these secretive creatures.

MORE FORESTS, FEWER SNOW LEOPARDS

The high mountains of central Asia are warming at three times the world average. As thick forests spread higher up the mountains, snow leopards will have less and less land on which to live. It is likely that by 2070, only one-third of the habitat snow leopards live in today will remain.

▲ Snow leopards thrive in high, rocky places where there are few trees.

WARMER MOUNTAINS, FEWER PREY

Warmer temperatures could also mean that farmers start to plant crops higher up mountains. They may keep livestock higher up the mountains, too. That would shrink snow leopard habitat further. It would also bring more farmers into contact with the leopards. Animals that snow leopards eat are adapted to cooler places, too. They will move higher up mountains as temperatures rise. There are fewer plants in higher places for those animals to eat, so there will be fewer prey animals for snow leopards.

▲ If snow leopard prey animals are forced to move to higher ground, they may not find enough to eat. Without food, they will become weak and die.

PEOPLE ON A MISSION: TACKLING CLIMATE CHANGE

A climatologist is a scientist who studies the climate. Climatologists look at the ways the world's weather has changed over long periods of time. They consider how those changes are affecting the planet, people, and wildlife. Climatologists examine **samples** of ice taken from deep beneath the polar regions to look for bubbles of gases trapped in ancient snow. That helps them learn about what the climate was like in the past. Their studies show Earth is getting much warmer today.

Some climatologists study a particular area, such as the high mountains where snow leopards live. They use **satellite** images to look at how much snow covered the mountains in the past compared with today. They compare glacier sizes in the past with the sizes today, so they can see how climate change is affecting them. A glacier is a large area of ice on land.

A climatologist can use the information they collect to create climate models using computer software. The models can be used to figure out what may happen to places such as snowy mountains in the future. For example, the scientists use computer models to figure out how the snow leopard's habitat is affected by the tree line moving higher. The scientists tell conservationists about their findings, so they can take action.

SNOW LEOPARDS NEED OPEN SPACES

The body of a snow leopard is perfectly designed for the mountains on which it lives. In this environment the powerful cat is a skilled and deadly hunter.

HIDE AND SEEK!

Snow leopards have the perfect **camouflage** for their habitat. Their patterned fur hides them against rocks and snow. That helps the snow leopards to sneak up on prey.

CLEAR VIEWS

Snow leopards can see far and wide over open, rocky spaces. That gives them clear views to spot prey. It also helps them look out for danger.

LONG-DISTANCE LEOPARDS

Snow leopards cover long distances over open ground in a short period of time. They are adapted to leap, bound, and pounce over rocks and ravines.

WRITTEN ON THE ROCKS

A snow leopard sprays urine to tell other snow leopards to keep away from its territory. The smell also sends messages to other snow leopards that are looking for a **mate**.

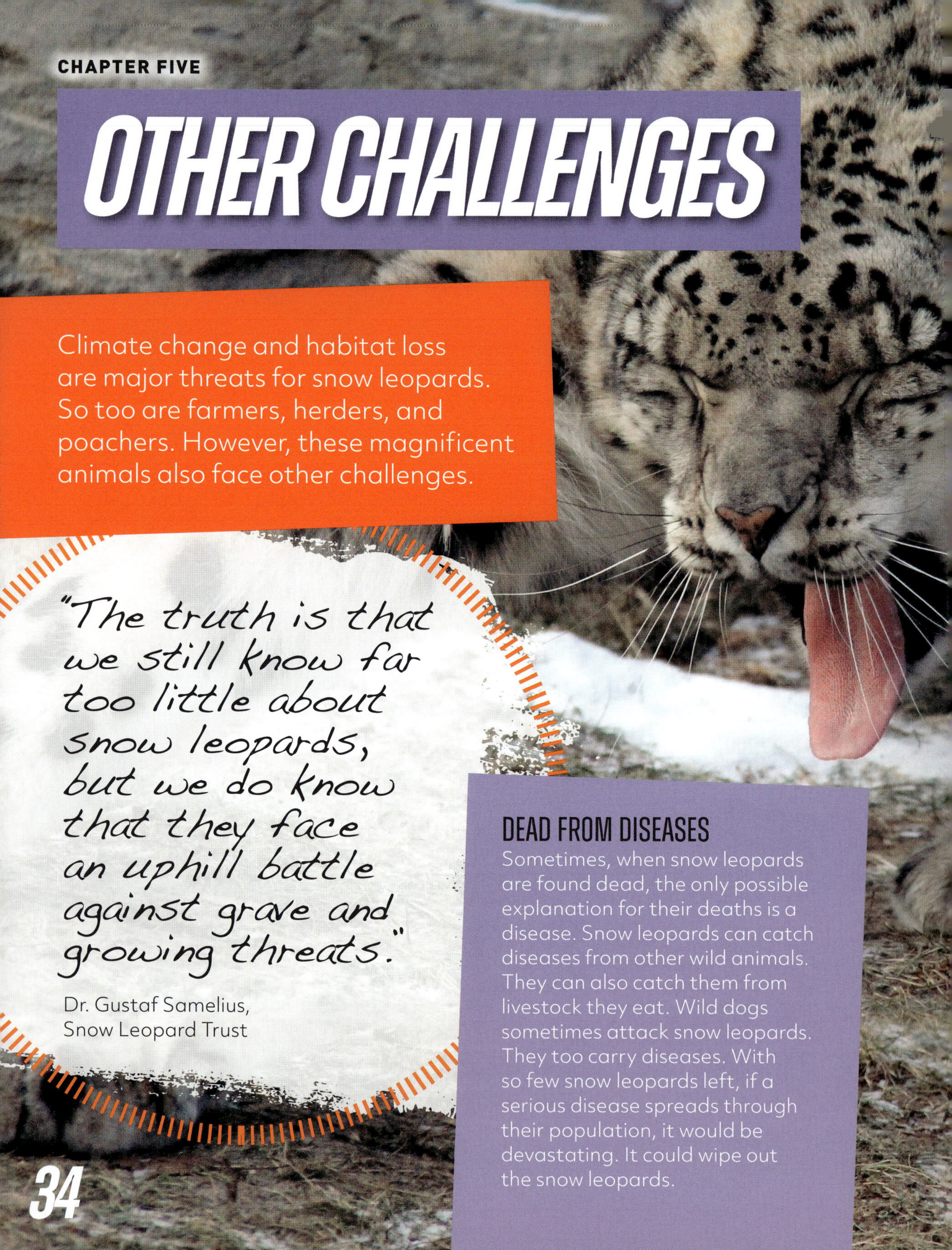

CHAPTER FIVE

OTHER CHALLENGES

Climate change and habitat loss are major threats for snow leopards. So too are farmers, herders, and poachers. However, these magnificent animals also face other challenges.

"The truth is that we still know far too little about snow leopards, but we do know that they face an uphill battle against grave and growing threats."

Dr. Gustaf Samelius,
Snow Leopard Trust

DEAD FROM DISEASES

Sometimes, when snow leopards are found dead, the only possible explanation for their deaths is a disease. Snow leopards can catch diseases from other wild animals. They can also catch them from livestock they eat. Wild dogs sometimes attack snow leopards. They too carry diseases. With so few snow leopards left, if a serious disease spreads through their population, it would be devastating. It could wipe out the snow leopards.

NOT READY FOR THE WILD

Zoos that breed snow leopards in **captivity** do so in the hope that some of these animals could one day be released into the wild. Unfortunately, at the moment, it is not possible to do that. Studies on other big cats suggest that less than one-third of zoo-bred snow leopards would survive if released into the wild. The problem is that animals born in zoos have less fear and take more risks than wild animals. They are therefore more likely to be killed. They also lack the hunting skills or disease **immunity** needed to survive in the wild.

◄ Diseases are a serious problem for snow leopards, especially with so few left in the wild today.

@eco_kofi_001

I wrote letters to several governments in countries where snow leopards live. I asked them to find out more about snow leopards there. I got a lot of people to sign the letters, too.

KNOWING TOO LITTLE

A major challenge facing people who want to help snow leopards is the animals' habitat. Snow leopards live in remote places. For that reason, researchers have reached only 23 percent of their habitat. That means there are huge holes in our understanding of what snow leopards need to survive and thrive.

SPOTLIGHT ON CAPTIVITY

Keeping snow leopards in captivity helps protect them. The animals that live in zoos are safe from threats such as habitat loss, angry farmers, poachers, and climate change. It is difficult to study snow leopards in the wild, because they are so secretive. It is much easier to study them in zoos. Scientists can then use the information they learn to advise conservationists how to help wild snow leopards.

GOOD NEWS FOR SNOW LEOPARDS

We cannot yet release snow leopards that are born in zoos into the wild. But the good news is that snow leopards breed, or produce young, well in captivity. Cubs born in captivity have a better chance of growing into adults than those born in the wild. In zoos, they are safe from predators or poachers. By breeding snow leopards, we can help keep the species from dying out.

▼ The hope is that, one day, snow leopards born in captivity may be able to live in the wild.

▼ Seeing snow leopards in zoos encourages people to care about them and learn more about how to help these animals.

PEOPLE ON A MISSION: SNOW LEOPARD SCIENTISTS

Zoologists study snow leopards to find out more about them, such as what they eat and how they raise their young. They study what diseases the animals suffer from.

Wild snow leopards are difficult to observe. That is why some conservation **geneticists** use **genetics** to discover more detailed information about them. These scientists examine animal **deoxyribonucleic acid (DNA)**. Animal DNA is a chemical that carries genetic information. It has all the instructions that a living organism needs to grow, make babies, and live. Scientists get the DNA from snow leopard scat, or poop, and hairs that were collected from the wild. Then scientists in a lab extract, or remove, DNA from the scat. They study it to find the genetic makeup of each individual animal. That can help scientists figure out exactly how many snow leopards live in an area and how they are related, even if no one has ever seen one there.

OTHER BIG CATS IN DANGER

Big cats around the world are under threat. Humans are taking actions that harm these beautiful animals or their habitats. If we are not careful, one day they may be gone forever.

THE CHEETAH

The cheetah is the fastest land animal in the world. Its tail is an unusual flat shape to help this big cat keep its balance when running very fast. Sadly, cheetahs cannot outrun a bullet, and they are endangered because of hunting. They are also at risk because their **grassland** habitats in Africa are being taken over by humans and destroyed by climate change.

THE JAGUAR

The jaguar is the third-biggest cat in the world, after the tiger and the lion. Jaguars look similar to leopards, except they have black dots in the middle of some of their rosettes. They also have larger, rounded heads and short legs. Jaguars are mainly found in South America. They are under threat because the Amazon rain forest where they live is being cut down. Jaguars are also hunted by humans.

THE LION

African lions are adapted for life in grassland habitats. They are at risk because farmers have taken over a lot of their land. The lions' grasslands are also threatened by climate change. Many lions have been killed by poachers, too.

THE TIGER

Tigers are the world's biggest cat. They are fearsome killers. Today, most tigers are found in India and Thailand. A few still live in Siberia and countries in Southeast Asia. These cats are killed for their fur and bones. The forests and grasslands where tigers live have been also destroyed to make room for homes, mines, and farms. Climate change is damaging the cat's habitat, too.

WHAT'S NEXT FOR THE SNOW LEOPARD?

No one knows for sure what the future holds for snow leopards. The animals and their habitat face many threats. However, around the world, conservationists and governments are working hard to ensure the survival of this rare and beautiful beast. So too are ordinary people.

HELP FROM CONSERVATIONISTS

As the world's population grows, more people are living near snow leopard habitats. Conservationists need to find ways for snow leopards and people to live safely near each other. One way to do that is by educating people. Conservationists explain why snow leopards are important to the ecosystem. They explain why they need to be protected. They work with local people to reduce poaching and farming.

Conservationists are developing new technology to help them monitor snow leopard populations. They are working with governments and other organizations to create more reserves. They are working hard to create more wildlife corridors. They are working to bring in stronger laws to stop the illegal wildlife trade.

@cali_tina_47

I'm studying hard in science, so that one day I can work in a lab. I want to use my skills to help snow leopards survive.

HELP FROM ORGANIZATIONS

The WWF and the Snow Leopard Conservancy Trust are two of the organizations fighting to save snow leopards. They help farmers build snow leopard-proof livestock corrals. They help families set up tourist homestays to give farming families a new way of earning money. WWF has also set up compensation schemes. They are systems that pay money to farmers and herders for livestock killed by snow leopards. That reduces the risk of farmers and herders killing snow leopards in anger.

HELP FROM ALL OF US

Everyone, everywhere can play their part in helping to save the snow leopard. We can all live, travel, and shop in more sustainable ways to reduce the threat of climate change. We can learn more about snow leopard threats, so we can raise awareness of the issues. By working together, it is possible to save the snow leopard. Then, this amazing king of the snow mountains will roam freely and safely across its habitat for many years to come.

MAKE IT YOUR MISSION: A CAREER IN CONSERVATION

Why not make it your mission to build a future career in conservation? That way, you could make a huge difference to snow leopards and other endangered animals. On pages 44–45, you'll find some of the conservation careers you could pursue. And here are things you can do right now to prepare for a career in conservation and help save precious snow leopards.

GET SET TO STUDY

Working hard at school now will pay off if you want a conservation career. Pay attention in your science, English, and geography classes. Languages are also useful. If you have a job oversees, being able to speak a foreign language is helpful.

GET CONNECTED

Connecting with other people who care about wildlife is a great way to feel empowered about saving endangered animals. You'll find some of the organizations you can join on page 47. By connecting with other people who are concerned about the environment, you'll find out what you can do to help. Being involved with conservation organizations and charities now will also help you when you apply for jobs in the future. It will show that you have been interested in conservation from an early age.

TAKE ACTION!

You have learned about some of the actions that help snow leopards from the Kids on a Mission features in this book. Here are some more ideas for other activities that can help protect snow leopards:

- Adopt a snow leopard! Ask your parents to help you check out the WWF site for information about adopting snow leopards.
- Buy products from snow leopard charity online gift stores for friends and families.
- Ask your teacher to teach your class about snow leopards. Some snow leopard charities have education packs that the teacher could use.
- Take action to stop climate change, from car sharing to get around, to turning lights off when you leave a room. Every little thing you do helps.
- Ask your school to hold a fundraising event to raise money for an organization such as WWF to help them protect snow leopards.
- Start a blog about snow leopards and the dangers they face. Educate as many people as you can.

▲ Learn all you can about the snow leopard and other endangered animals. The more you learn, the more you can educate others about Earth's threatened species.

GET EXPERIENCE

Any form of work experience in wildlife care will be useful for a future conservation career. Why not apply to local zoos, charities, and conservation organizations as soon as you are old enough?

Cycling to school instead of ▶ traveling by car is a great way to take action to stop climate change. It's also fun and keeps you fit.

CAREERS IN CONSERVATION

More and more people are becoming concerned about the environment and what the future holds for the world's wildlife. Many of them are looking for jobs in conservation. If you would like to pursue a career in conservation, here are some of the jobs you could choose. You can find out about other conservation careers on page 47 of this book.

▲ Wildlife biologists spend a lot of time studying animals and their environments. That helps them learn what the animals need to survive.

WILDLIFE MANAGER

If you became a wildlife manager you would check on the health and well-being of animals in a natural area. You would also take care of the animals' habitat. Another job of a wildlife manager is educating the public about animals and the need to protect them. An environmental studies degree is usually needed.

WILDLIFE LAW ENFORCEMENT OFFICER

The job of a wildlife law enforcement officer is to help protect wildlife. You would make sure that people follow wildlife laws and rules. You would also conduct surveys of wildlife populations. You might be involved in trapping and **tagging** programs for animal surveys, too. A wildlife law enforcement officer also follows up complaints about wildlife that is a nuisance to farmers or local people. Educating local people about wildlife is another part of the job.

WILDLIFE BIOLOGIST

If you are good at science, then you might consider a job as a wildlife biologist. A wildlife biologist collects and studies data on wildlife and habitats. This might include looking at animal behavior, diseases, genetics, food, or population. It may also involve studying the way things such as pollution affect wildlife and how to improve habitat conditions.

CLIMATOLOGIST

Climatologists are problem-solvers who may be expected to perform a variety of different tasks. As a climatologist you might conduct hands-on research, such as taking snow or soil samples. You would then analyze the samples and report on the data. Climatologists also present their research to other scientists at conferences.

JOBS IN ENVIRONMENTAL SCIENCE

There are many different roles in environmental science. For example, you could be a climatologist or a conservation scientist. A degree in environmental science is usually required for these jobs. If you want to become an environmental lawyer, you will need a law degree with a specialization in environmental law.

PUBLIC EDUCATION AND OUTREACH SPECIALIST

In this role you would teach others about wildlife and habitats. That could include people in areas far away from habitats and those who live nearby. Giving talks to children in schools to educate them about the importance of protecting wildlife is also part of the job. So too is giving talks to people in organizations, including companies interested in helping wildlife. A degree in environmental studies is needed for this role.

◀ Teaching people about wildlife and the threats animals face is a vitally important way of saving them.

GLOSSARY

advocacy getting support from other people to help you express your views

atmosphere the blanket of gases that surrounds Earth

bird of prey a bird that has a hooked bill and sharp talons for catching prey

camouflage patterns or colors that help animals blend in with the environment

captivity kept in an enclosed space, not in the wild

coexist live alongside

communications the sharing of messages or information

conservation protection of the planet

conservationists people who try to protect the planet

deoxyribonucleic acid (DNA) material that carries all the information about how a living thing will look and function

enclosures spaces with walls or fencing in which animals are kept

endangered facing a very high risk of extinction in the wild

environments natural places where plants and animals live

extinction dying out

food chain the animals and plants that are connected because they eat or are eaten by one another

fossil fuels fuels formed from the remains of plants and animals that lived long ago

geneticists scientists who study genetics

genetics related to the genes and the way in which characteristics are passed from parents to children

Geographic Information Systems (GIS) computer systems that analyze and display geographical information

geological related to Earth's structure

Global Positioning System (GPS) a system of satellites in space that tells us where something is

grassland large open area covered with grass and few trees

habitat a place in which plants and animals live

hydroelectric power a form of electricity made using the power of moving water

immunity ability to fight off diseases

mate an animal of the opposite sex to breed with

minerals substances found in rocks, such as gold and silver

mines openings in the ground from which resources such as minerals are taken

nurses feeds with milk

organisms living things such as plants and animals

policy a plan for action followed by a government or a group of people

pollute to make air, water, or soil dirty or harmful to living things

populations all the members of a species living in certain areas

precious metals metals that are valuable, such as gold or silver

predators animals that hunt and eat other animals

prey animals that are hunted and eaten by other animals

resources things that people use

rodents animals with large front teeth such as rats, mice, and squirrels

samples small amounts for testing

satellite a man-made object in space that circles Earth and sends back data

sea level the average level of the surface of the ocean

species a type of plant or animal

sustainable can be relied upon for the forseeable future

tagging attaching an electronic marker to an animal to track its movements

territory an area of land that an animal regards as its own and that it may defend from other animals

trade the exchange of goods for money

vulnerable facing a high risk of extinction in the wild

FIND OUT MORE

BOOKS

Adams, Sabrina. *Endangered Species and Our Future* (Spotlight On Our Future).
Rosen Publishing, 2022.

Markovics, Joyce. *Snow Leopards* (On the Trail: Studying Secretive Animals in the
Wild). Cherry Lake Publishing, 2021.

Mcghee, Karen. *World's Most Endangered Animals* (The World's Most).
Gareth Stevens Publishing, 2021.

Stuckey, Rachel. *Bringing Back the Snow Leopard* (Animals Back from the Brink).
Crabtree Publishing Company, 2019.

WEBSITES

Discover more about careers that help fight climate change at:
www.bestcolleges.com/blog/climate-change-jobs

Discover the ultimate guide to careers in conservation at:
www.conservation-careers.com/15-key-conservation-jobs-ultimate-guide-for-conservation-job-seekers

Hear directly from people working in conservation. Find out what they have to say
about a career in conservation at:
www.conservation-careers.com/conservation-jobs-careers-advice/how-to-get-a-job-in-conservation

Find out more about a fascinating career in wildlife forensics at:
www.environmentalscience.org/career/wildlife-forensics

Discover careers in environmental science at:
https://jobs.environmentalscience.org

Find lots of amazing wildlife careers and what they involve at this useful site:
www.thebalancecareers.com/careers-with-wildlife-125918

Publisher's note to educators and parents:
All the websites featured above have been carefully reviewed to ensure that they are suitable
for students. However, many websites change often, and we cannot guarantee that a site's
future contents will continue to meet our high standards of educational value. Please be
advised that students should be closely monitored whenever they access the Internet.

ABOUT THE AUTHOR

Award-winning author Louise Spilsbury, who also writes under the name Louise Kay Stewart, has written more than 250 books for young people on a wide range of subjects. She especially loves writing about animals and learning more about what we can all do to protect amazing species such as the snow leopard.